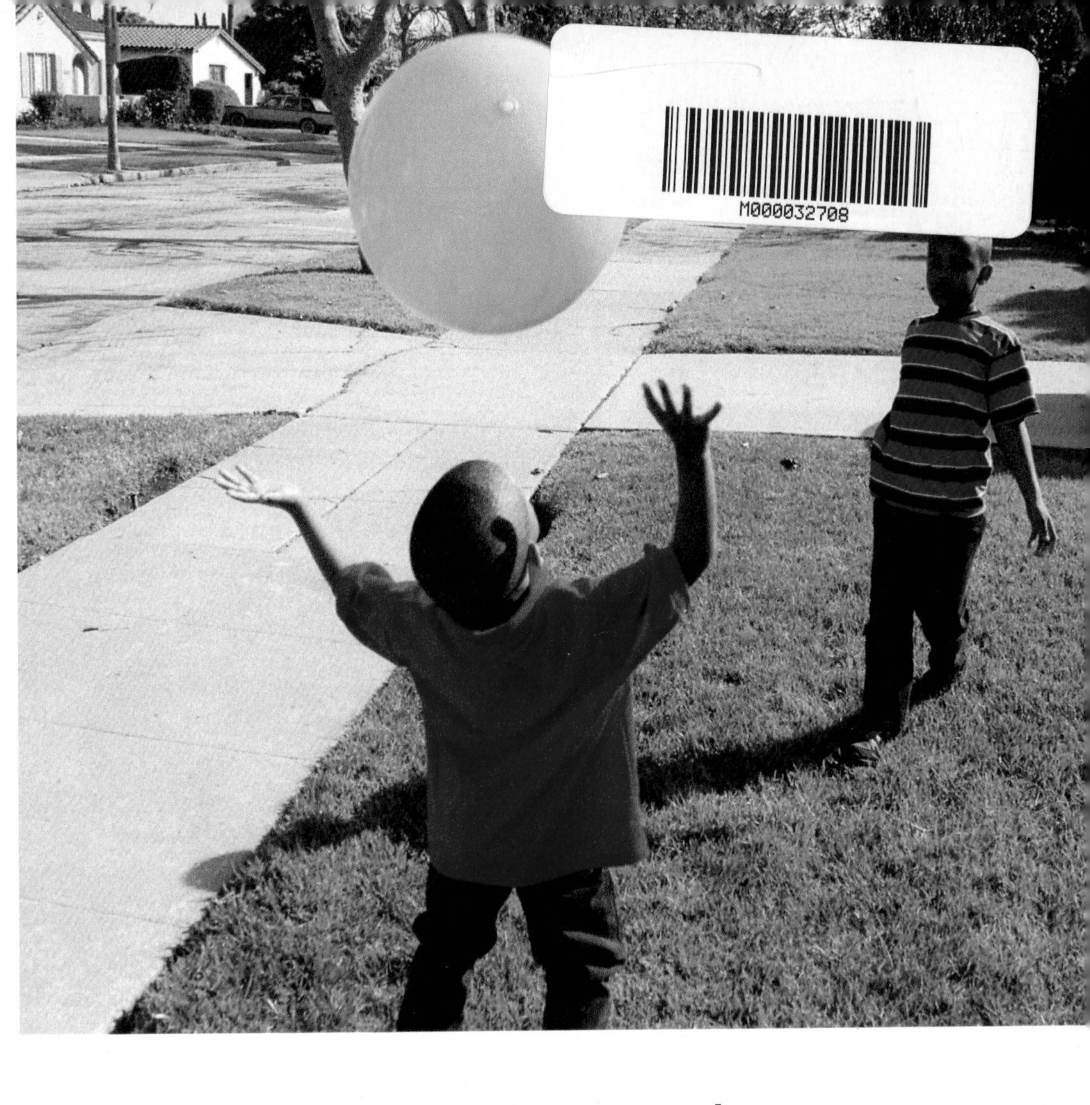


This ball is a sphere.

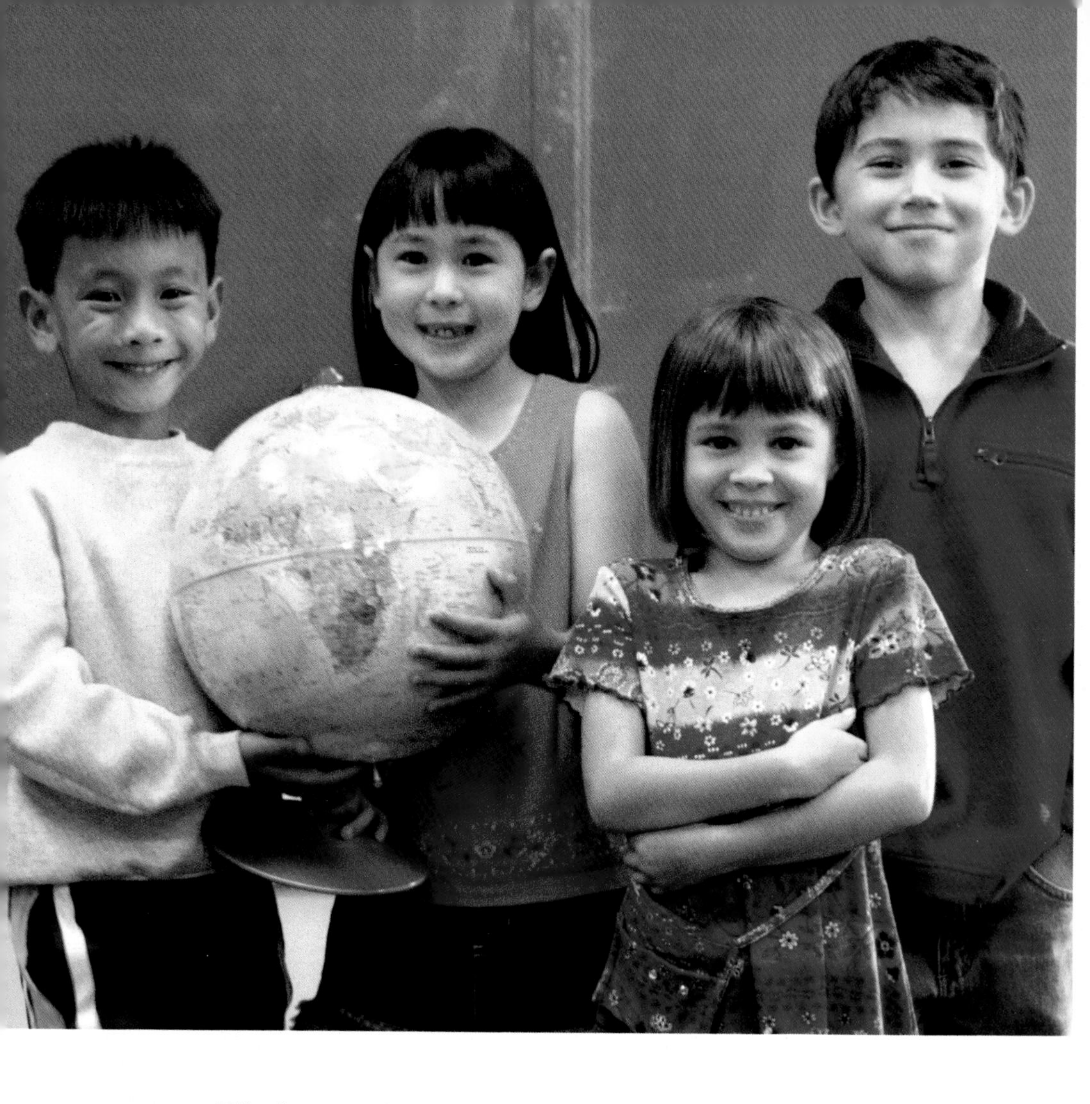

This globe is a sphere.

# Sphere

by Jennifer Boothroyd

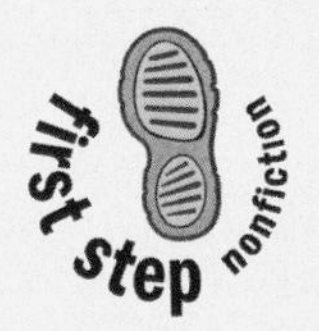


Lerner Publications · Minneapolis

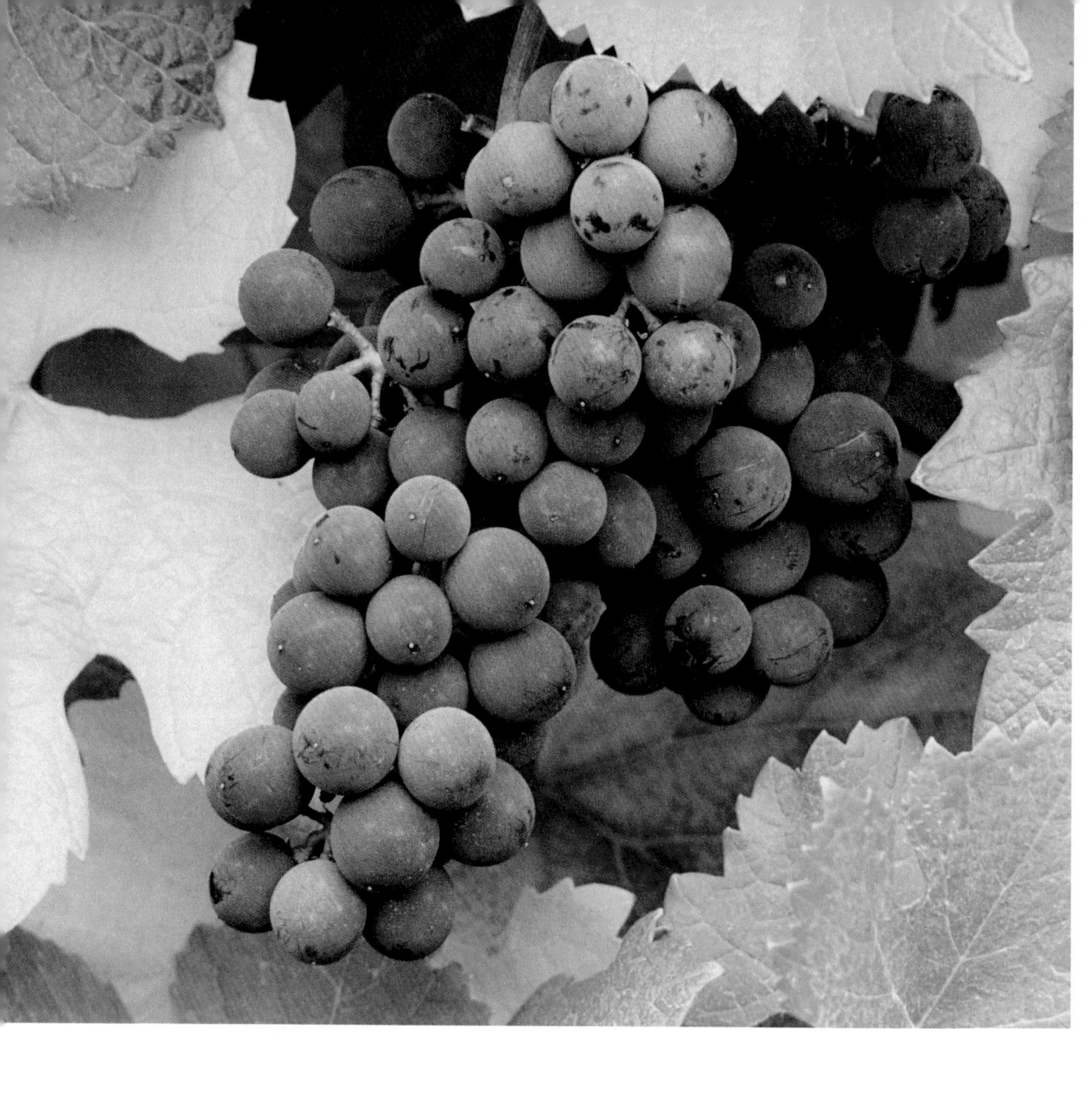

I see a sphere.

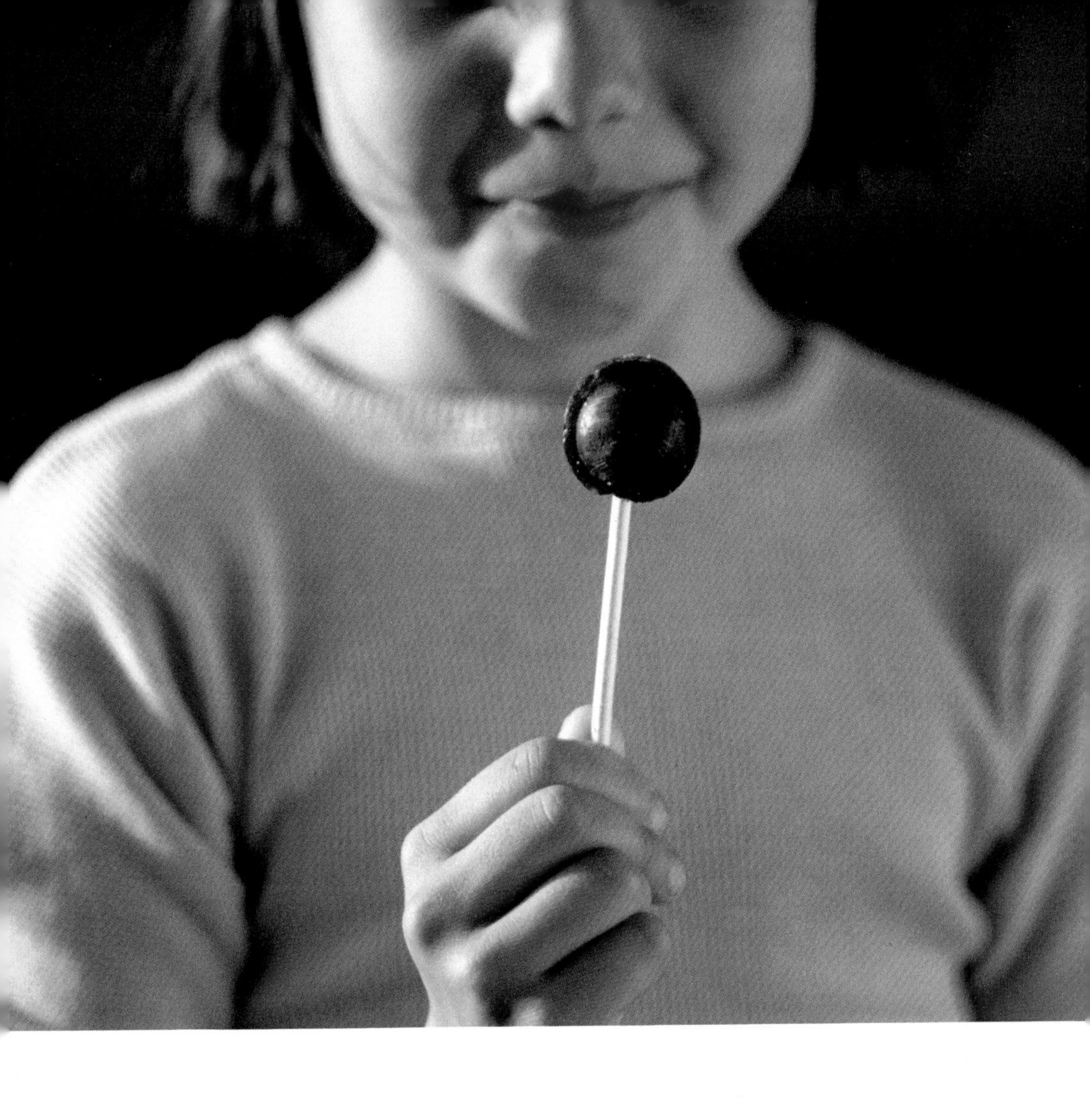

This candy is a sphere.

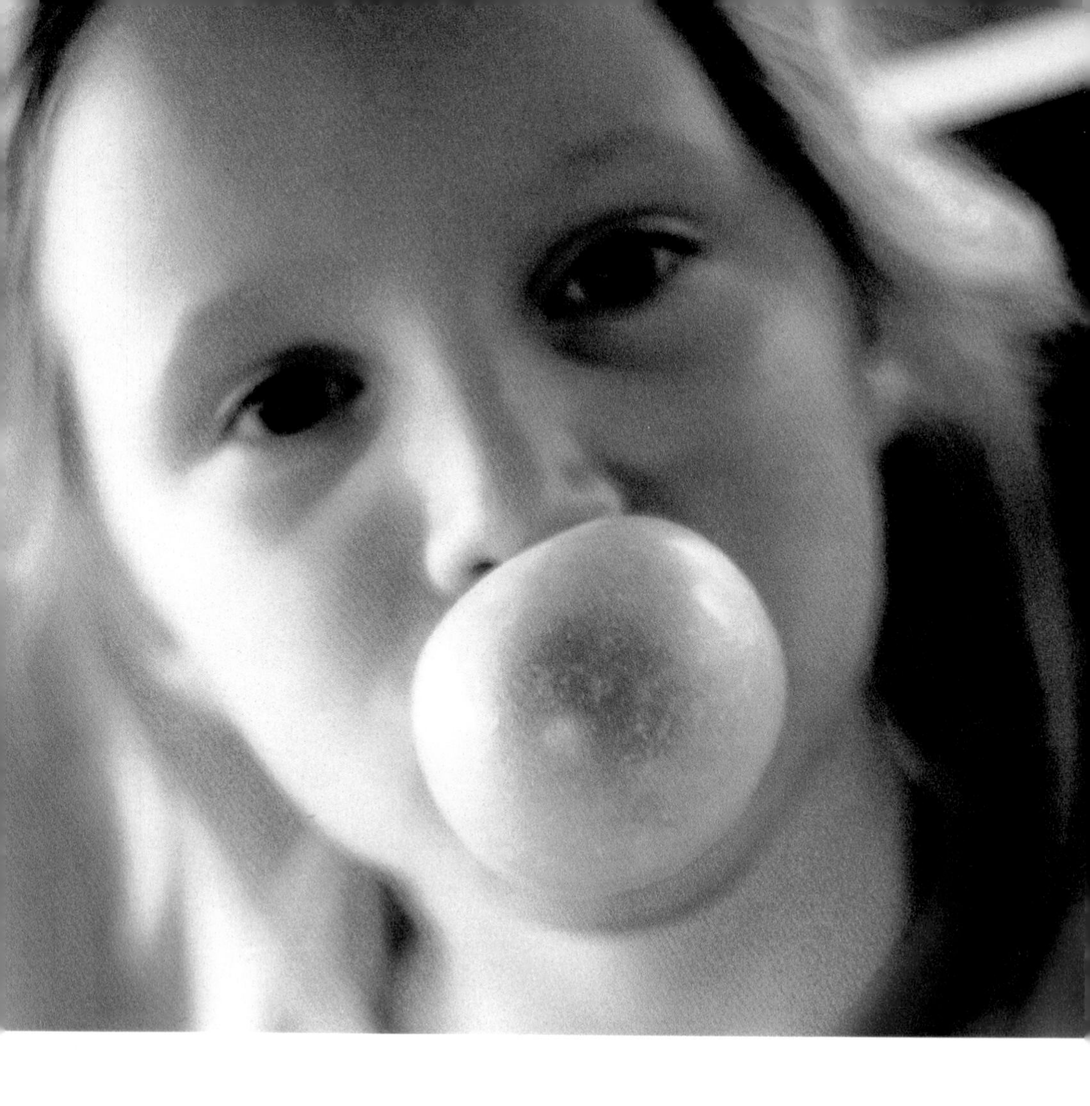

This bubble is a sphere.

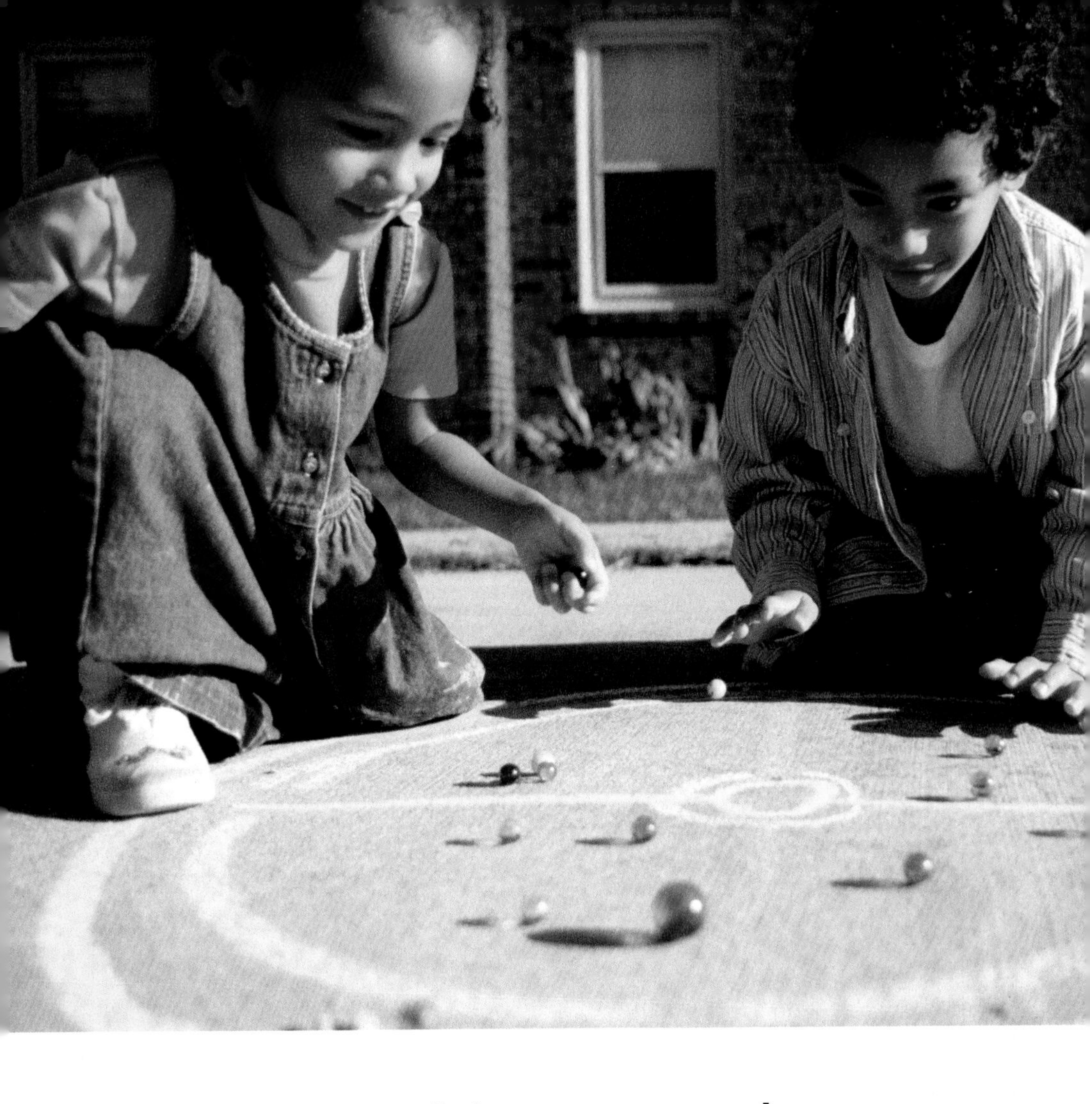

These marbles are spheres.

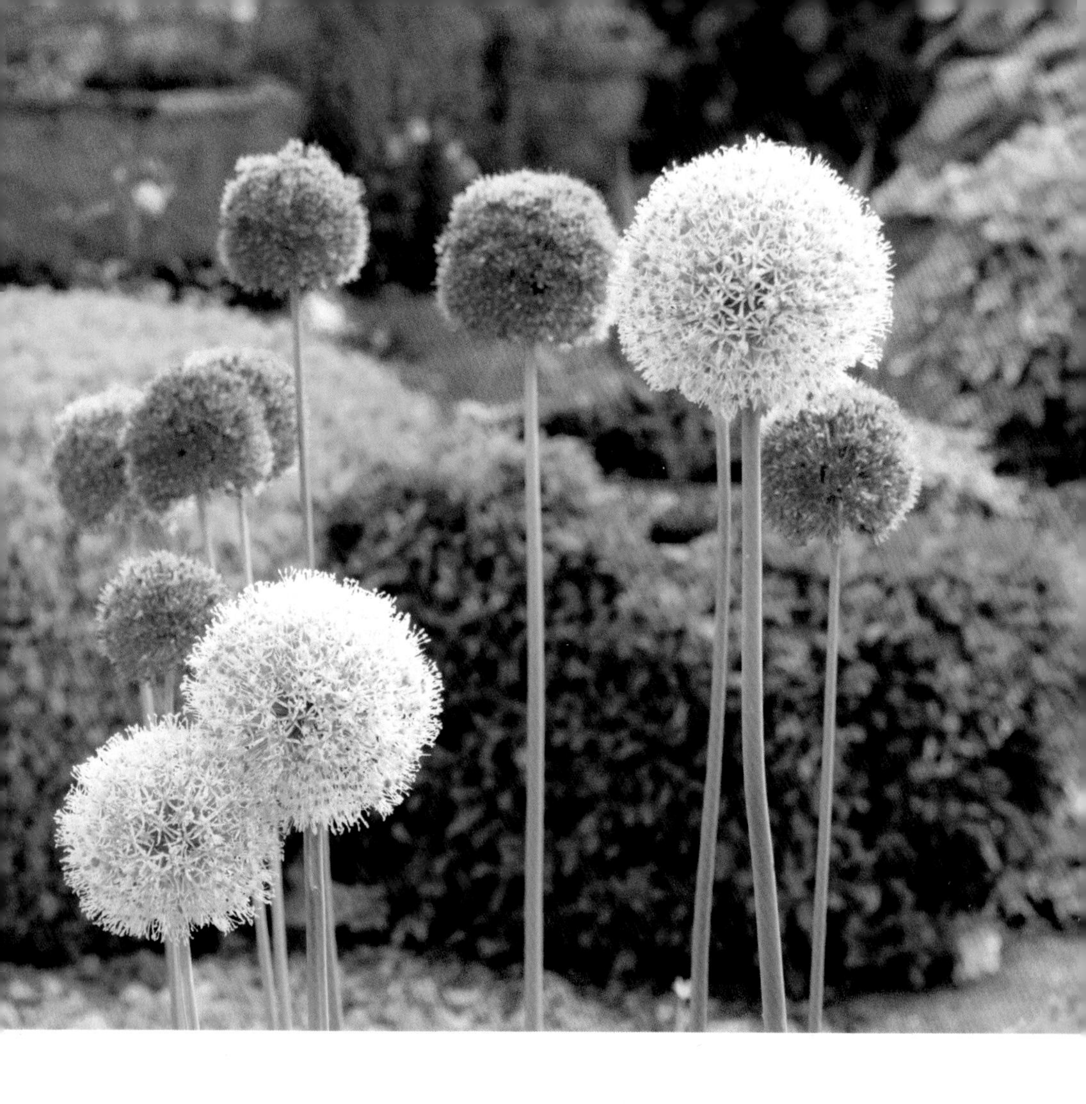

Do you see spheres?